Landforms

Valleys

Cassie Mayer

Heinemann Library
Chicago, Illinois

Photo research by Tracey Engel and Tracy Cummins
Designed by Jo Hinton-Malivoire
Printed and bound in China
16 15 14
10 9 8 7

Library of Congress Cataloging-in-Publication Data
Mayer, Cassie.
 Valleys / Cassie Mayer.
   p. cm. — (Landforms)
 Includes bibliographical references and index.
 ISBN 1-4034-8436-8 (hc) — ISBN 1-4034-8442-2 (pb)
 ISBN 978-1-4034-8436-9 (hc) — ISBN 978-1-4034-8442-0 (pb)
 1. Valleys—Juvenile literature. I. Title. II.Series.
 GB562.M39 2007
 551.44'2—dc22
                        2006004672

Acknowledgments
The author and publisher are grateful to the following for permission to reproduce copyright material:
Alamy pp. **10** (Jon Arnold Images), **15** (Leslie Garland Picture Library); Corbis pp. **4** (mountain, Royalty Free; volcano, Galen Rowell; island, George Steinmetz), **5** (Pat O'Hara), **6** (Richard Klune), **7** (Pat O'Hara), **11** (Jon Sparks), **12** (Keren Su), **13** (Franz Marc Frei), **14, 16** (Royalty Free), **17** (Dean Conger), **19** (Pablo Corral Vega), **21** (Ashley Cooper), **22** (river, Royalty-Free; rhino, Alissa Crandall; village, Keren Su); Getty Images pp. **8** (PhotoDisc), **9** (PhotoDisc), **18** (John Lawrence), **20** (Kate Thompson).

Cover photograph of clouds over the Aniscio Canyon reproduced with permission of Corbis/Francesc Muntada. Backcover image of the Hooker Valley, New Zealand reproduced with permission of Corbis/Franz Marc Frei.

Every effort has been made to contact copyright holders of any material reproduced in this book.
Any omissions will be rectified in subsequent printings if notice is given to the publisher.

Contents

Landforms

The land is made of different shapes.
These shapes are called landforms.

valley

A valley is a landform.
A valley is not living.

What Is a Valley?

A valley is a low piece of land.

Valleys are between mountains
and hills.

Valleys form over many years.
Water runs down a mountain.

Water forms a river.
The river creates a valley.

Valleys change shape over time.

Wind and water wear valleys away.

Types of Valleys

Some valleys are shaped like a "V."

This valley is shaped like a "V."
It was formed by a river.

Some valleys are shaped like a "U."

This valley is shaped like a "U."
It was formed by a river of ice.

Some valleys have steep sides.

Some valleys are filled with water.

What Lives in a Valley?

Valleys are home to living things.

Animals live in valleys.
People grow plants and food in valleys.

Visiting Valleys

People like to visit valleys.

Valleys show us the highs and lows of the land.

Valley Facts

A canyon is a type of valley. The Grand Canyon is in Arizona. It is one of the largest canyons in the world.

The Great Rift Valley is in Africa. It is home to many different animals.

Picture Glossary

steep almost up and down

Index

Note to Parents and Teachers

This series introduces children to the concept of landforms as features that make up the earth's surface. Discuss with children landforms they are already familiar with, pointing out different landforms that exist in the area in which they live.

In this book, children explore the characteristics of valleys. The photographs draw children in to the natural beauty of valleys and support the concepts presented in the text. The text has been chosen with the advice of a literacy expert to enable beginning readers success reading independently or with moderate support. An expert in the field of geology was consulted to ensure accurate content. You can support children's nonfiction literacy skills by helping them use the table of contents, headings, picture glossary, and index.